EMERGENCY SOFTWARE SYSTEM

QUICK RESPONSE STYLES AND TECHNIQUES

DR. ABDULAZIZ Y. BAHHA

Outskirts Press, Inc.
Denver, Colorado

CONTENTS

LIST OF FIGURES

LIST OF TABLES

ACKNOWLEDGEMENTS

THIS FOCUSED ISSUE would not have been possible without the substantial support of a large constituency. I would like to thank Dr. Stephen A. Szygenda, *Professor of the Department* of Engineering Management, Information, and Systems and Department of Computer Science and Engineering at SMU, for assisting me with his knowledge and feedback. I am grateful to him for his wisdom, guidance, and opportunities. My thanks also goes to other members of my committee, Dr. Jerrell Stracener, Dr. Jeff Tian, Dr. Frank Coyle, and Dr. LiGuo Huang, for reviewing the manuscript and making guiding suggestions. I am very grateful for opportunities and support offered by Collin County Homeland Security. I have benefited greatly from discussions with FBI Agents Mr. Stone Kelley and Mr. Gamal Abdel-Hafiz.

I would like to mention that the librarians were gracious enough to provide multiple ways for me to contact them via: Email, Chat, Visit, Phone, Appointment, and special services for businesses. Special thanks goes to Ms. Cassie Bone for reviewing the English language using her knowledge of spelling, grammar, and proofreading. Ms. Bone helped me find balance in life despite the nights and weekends that I spent hunched over a keyboard. Additional thanks to the many people that researched the information that provided the textbooks such as Interaction Design, and Software Requirements: Styles & Techniques and Requirements Engineering. Finally, many thanks to my family and especially to my colleagues, Mrs. Karen Hanvey and

Ms Beckey Hargett both of whom provided particularly helpful suggestions and recommendations.

The purpose of the "An Emergency Software System for Quick Response" is to effectively operate the whole emergency system so that any incident or accident can be dealt with in a quick response and limited delay. The system is a computer based system so that we can achieve speed as well as accuracy in police work. At present the processes in the emergency system are manual. Some of the processes are mentioned below:

Whenever an incident occurs involving only a policeman or an informer, information about the incident goes to different police stations. Hence, there is no problem in communication and of course action. It is very difficult to track down the vehicle available and its position. It is a paper based system, so more time is needed to find out pertinent information. In spite of that, the information cannot be shared on a regular basis. The same problem is happening with tracking the special vehicle available at a different police station. When the incident occurs the policemen are not getting proper information about the availability of the special vehicles and machines.

The current manual system does not have a specific knowledge about the locations and the problems for going there. If an incident occurs sometimes the police follow a long route or busy traffic route, hence the delay in handling the case.

The central administration that controls every police station has less technical machines to get updated information.

To override the above problems occurring due to a manual system a computer based system will be designed that will overcome the manual delay as well as making the whole system efficient. The following users will be using the system:

- Police operating in different parts of the city.
- Police Administrators for tracking the operation of the incidents.

- Emergency Medical Services (EMS) to update information about the ambulance vehicle availability.
- Fire Department for information about the current and future requirements and non emergency personnel for informing about the incidents and happenings.

The system will be placed at various police booths where the operator will work on the system 24x7 to provide the updated information to other police. Apart from that the system will be used at an administrative level for tracking the activities at various police stations. The following factors will be helpful to the economics of the whole system:

We will use the technology that will be platform independent and will have little or no setup cost. The user interface for this system will be so simple that the time of training required will be minimal. The system will use less hardware and give better output.

1

INTRODUCTION

1.1 Emergency Services

BEFORE THE 1970s, Emergency Services used a manual process to dispatch emergency vehicles. The communications center received calls from citizens, recording the necessary information on paper forms, or call logs. Using a map on the wall, the operators attempted to place a pushpin to mark each spot. In the 1970s, calls between a dispatcher and the emergency vehicle was still problematic. The dispatcher had to call each separate agency needed at the scene. The system did not permit all emergency vehicles to be contacted at the same time.

According to Kelly Stone of the Collin County, Texas, head of Homeland Security and FBI agent, Computer Aided Dispatch (CAD) software was used to send messages to the dispatcher via a Mobile Data Terminal (MDT) and/or used to store and retrieve data such as radio logs, field interviews, and citizen information. If the dispatcher entered a location that the computer could not find in its Geodatabase (Geographical file to organize and store digital images), the software would display a list of close matches from which the dispatcher could select the correct location. So, if no exact match was found, the system would select the closest match. A dispatcher may announce the call details to field units over a two way radio between call centers and vehicle services using a slow communication service. The internet was not widely available. In the time that it took to relay those messages, lives were potentially lost.

In the 1990s, the system improved to where the call center and vehicle services were put in touch, but the issue still remained of having to connect to each service individually. By the 2000s, computers were found in a majority of households. Software costs were affordable for companies. According to Dispatch Magazine On-Line, in 2003, statistics found that just 33% of all local emergency service departments used computers for dispatching. The percentage is skewed by low numbers of computers by agencies for service populations under 10,000.

In order to improve the emergency services software industry problem of too many independent factions using different systems, if a system was even used this paper will offer a streamlined solution. The solution is to have six regionally positioned 911 call centers throughout the United States that used one portal based on one software system that all emergency services personnel may access. The figure 1.1 illustrates the six regions that divide the country. The 911 operators would receive all calls from the states within those areas. Cities and towns would be charged a percentage of overhead and labor. Since these costs are shared across the region, cities and towns would not spend as much money from the annual budget.

The software system contains different types of technology that will be detailed in the following paragraphs:

- The software allows operators to track emergency vehicles. The communications center receives calls from citizens and records the necessary information to collect, manage, and report assorted data.
- Many of these software systems are enhanced with Geographic Information System (GIS) capabilities and integrated with vehicle-based technologies such as Mobile Data Terminals (MDTs) and Automatic Vehicle Location (AVL) (FTA, 2007).
- Emergency services software allows data to be shared on the street through Mobile Data Terminal (MDT) PC.

- Computer Aided Dispatch (CAD) is usually integrated with automatic vehicle location (AVL). Geographic Information System (GIS) will formulate statistical analysis on the information that was collected from the regional centers weekly, monthly and annually. Mobile Data Terminal (MDT) is a computerized device used in emergency vehicles to communicate with a central dispatch office.
- Automatic vehicle location (AVL) is a computer-based vehicle tracking system. The actual real-time position of each vehicle is determined and relayed to a control center. Other advanced system features are often incorporated with AVL systems (FTA, 2007).

Emergency services software was implemented over three stages.

1. In the first stage, radio communications equipment will be upgraded to improve communications between emergency vehicle operators and dispatch.
2. The second stage of the project will consist of using Computer Aided Dispatch (CAD), scheduling and billing software to increase the efficiency of vehicle usage, and cost allocation administration.
3. During the third stage mobile Global Positioning satellite units will be installed in emergency vehicles so that the Automatic Vehicle Location (AVL) system can provide the dispatcher with real-time vehicle location to assist in routing.

The main work in Emergency Services Software:

The combination of AVL, GPS, CAD system and GIS is the catalyst to dispatch the closest emergency services branch to the incident scene.

System objectives statement:

Mobile computing tools can be used to reduce power outage duration, manage field crews, improve service scheduling, capture field data at the source, improving accuracy, and reducing operating costs (Waldman, 1999).

Original work for the software system:

The United States will be divided into six areas with an emergency services center in each region. All centers use the same software and coding. There are six centers providing visibility for the citizens that are in close proximity. The figure 1.1 below illustrates the six regions (The Northeast, the Middle Atlantic, the South, the Midwest, the Southwest, and the West) that divide the country, and how the various emergency service software system elements can work together in emergency vehicles to assist the citizen, operator, and emergency vehicle. The nation is divided into six regions because it is easier to manager (geographical proximity) Metrics for achieving objectives in the software system:

The measure for achieving the objectives is by using incident response times. This has long been the metric for emergency services software systems to evaluate the quality of service delivered to the citizen. There is a relationship between incident response times and incident outcomes. Incidents with the shortest response times result in the best outcomes, when the incident involves fire, an emergency medical situation, or accident.

The figure 1.1 Illustrates the Six Regions that Divide the Country

1.2 Implementation of a GPS Based AVL Emergency Software System in Regional Center

An emergency operations center provides a central point of control and coordination during major emergency situations throughout the six regions. Command center personnel must synthesize information and make rapid decisions as events unfold in real time. It is essential that personnel at an emergency operations center share a common operating perspective (ESRI, 2005). The implementation of a GPS based AVL system for six regional emergency software systems has been presented in this paper. This system is composed of four main parts: CAD, MDT, digital communications and fuzzy logic. A fuzzy logic system helps capture knowledge and relationships that are not precise or exact such as misinformation (wrong address or spelling of names). For example, if a map matching method was implemented for software, which uses a digital road map to correct the GPS position error, the software would display a list of close matches from which the dispatcher could select the correct location. If no matches were found, the system will select the closest match (Jacobson, 1999).

The software is equipped with many capabilities which can be classified into two main categories (Guo, Ji, & Hu, 2002):

1. The first category is implemented as a Graphic User Interface (GUI) to make the user interface design friendly. The instantaneous position of a mobile vehicle on the map with zooming capability and the momentary geographical position of the mobile vehicle can be viewed graphically. Factors such as date and time, velocity, error factors, and channels status of a GPS receiver can be extracted for each mobile vehicle.

2. The second category of software capabilities is searching and path-finding. The best path for a mobile vehicle to a desired destination can be found instantaneously based on determined criteria. The implemented AVL system can be in a state, city, or province.

According to Kelly Stone of the Collin County, Texas, head of Homeland Security and FBI agent, for transit, the actual real-time position of each emergency vehicle is determined and relayed to a dispatch center. The basic ideas of AVL systems in Emergency Software System in Quick Response include: computer-aided dispatch software, mobile data terminals, emergency alarms, and digital communications. The GPS map for emergency services exists in digital maps, GPS map helping tools errors is not possible to ensure that mobile positions register properly on a digital map. To avoid this problem, the map matching method can be used to improve mobile position display precision over an electronic map. A map matching method using a digital road map is a useful approach to correct these errors. There are a lot of maps matching methods based on GPS. The new idea in this paper, using fuzzy logic matching algorithm is proposed for reducing a GPS position error in an AVL application. It is easy to implement and requires little computation. Also, this method can find an exact mobile position on the road. Implemented software of the GPS based AVL system exploits this method for precise map matching. The effectiveness and applicability of the proposed algorithm is verified by the real road experiments and help both the emergency vehicle operator and the dispatcher (Mosavi & Ghadiri, 2002).

1.3 Integration Location Applications

This proposal illustrates the six regions that divide the country. When a citizen calls the dispatch center, the GPS, CAD, and AVL integrated extended applications are associated with emergency vehicle tracking and routing. The dispatcher will check if an emergency vehicle is available by using the AVL received, then a GPS signal updates the set location every few seconds sharing data with the CAD system. Each emergency vehicle has AVL to locate units that are available and close to the incident. The CAD tracks the location of all AVL providers with whatever is needed for use in a fuzzy logic database, which presents answers in probabilistic terms. The database contains all streets using the software that displays a list of close matches from which the dispatcher will select the correct location. Once an emergency incident is entered into the CAD system, the location is verified by the CAD streets database. The CAD system then recommends the dispatch center based on (Marsh, 2007):

1. The incident location such as state, city and street.
2. A predefined response plan according to incident type (low danger or high danger).
3. Which type of vehicle is needed and how many, (i.e.: police, EMS, fire truck).

This process is known as "dynamic dispatching", which is based on the updated GPS location. Dynamic dispatching represents a model shift from characteristics of the former method of emergency dispatching based on incident areas (Kuiper, 2000). The most effective use of integrated applications in an emergency incident location uses dynamic dispatching to reduce overall incident response times. It also allows all emergency vehicle operations to meet requirements for the emergency response goals of *The Standards of Response Coverage (2005)*. Emergency services installed on the AVL system along with mobile data terminals (MDT) system to be first use equipment. Integration of AVL with the CADs was intended to optimize system resource allocation across the jurisdiction, even when units were out of their assigned areas.

The purpose of the AVL was to provide the real time GPS location in relation to the CAD system. The system enables dispatchers to assign the closest available resources to an emergency incident. The CAD system solution would be based on the actual position of available resources, and not based upon the dispatcher's best time *(2006-2010 Master Strategic Plan*, 2006).

1.4 The Solution

The solution is based on the assumption that all cities, towns, and municipalities will agree to allow six regional 911 call centers placed throughout the United States to represent them. The centers will have one portal based on one software system that all emergency services personnel may access. The results and analysis of the study will support documentation from system implementation. Cities, towns, and municipalities would be a charged a percentage of overhead and labor. Since these costs are shared across the region, cities and towns would not spend as much money from the annual budget. GIS will formulate statistical analysis on the information that was collected from the regional centers weekly, monthly, and annually. Those statistics are used to determine local budgets. The cities, towns, and municipalities will be able to see how much money will be saved after implementing the regional model. The impact of system effectiveness may facilitate documentation to satisfy emergency services reporting requirements.

According to Kelly Stone of Collin County, Texas, head of Homeland Security, and FBI agent, 2100+ vendors sell different software that tracks offense reports. The vendors use many programming languages to write the software, such as C++, Perl, and VB, C# or Java. This makes it difficult for cities, towns, and municipalities to access to multiple records. This problem would be solved by using one software system with the six regional centers. The reason that the United States has not implemented the six regional emergency services model before now is because the local cities, towns, and municipalities have the option of determining how to setup their current emergency services system and standard time zones (Gary Lum 2006). There are different ways for dispatchers to meet the requirements to work with the communications systems such as wireless internet and the vehicle information automated pursuit report. The product life cycle for the emergency system used the interaction software development model. Interaction is also referred to as interactivity. It is defined as a function of input required by the learner while responding to the computer and the analysis of those responses by the computer and the nature of the action by the computer (Sims, 1997).

1.5 Who is Responsible for System Development and Operation?

Assuming that all cities, towns, and municipalities will agree to allow six regional 911 call centers placed throughout the United States to represent them. There are several groups of people, such as architects, project managers, team leaders, software developers, and test engineers, who are responsible for the Emergency System's development, each having its own role. Figure 1.5 shows roles and responsibilities of the Emergency System (Booch, 2004).

The architects are responsible for building the emergency services system successfully. The main role of an architect is to manage the fundamental and advanced requirement of the system. He/she mediates between staff and the city manager. The architect is responsible for passing high-level information, such as formal English language, and breaking it down into a more technical level for the project manager.

The project manager's responsibility is scheduling and maintaining the project flow. He/she provides key business level information to the team leader, who, in turn, manages the team of developers. With the help of the team leader, the project manager breaks down the tasks and calculates resource allocation. He/she manages project flow, timeline, resource management, and risk management.

Team leaders are responsible for managing the development team. They look for the standards and quality given by the project manager. They also coordinate with team members to discuss the complexity and the issues of the codes. Team leaders also provide quality information to the project manager.

Software developers are the foundation of the roles for a software system. The software engineers are responsible for creating and developing modules, implementation of the modules as a code, and coordinating with team leader to achieve the project timeline. Developer codes are used to program the conceptual design in real time entities. Developers will decide which program languages will be used to write the software, such as C++, Perl, or Java. They translate computer science language into friendlier terminology via business logic (Booch, 2004).

The test engineers are responsible for creating test cases according to the flow and condition of the project. Test engineers are the parallel part of the whole project. They provide a "bug" list once the testing is over.

A project manager, or developer, is responsible for giving the test case to a test engineer. The test engineers are responsible for releasing the finished product. They will provide future updated releases of improved software over time.

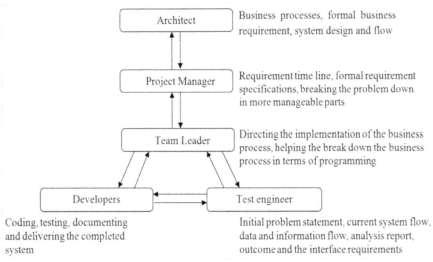

Figure 1.5
Roles and Responsibilities of Emergency System Personnel

2

DEFINITION OF PROBLEM

2.1. The Existing Emergency Services

THE EXISTING EMERGENCY services process for quick response uses a manual process to dispatch emergency vehicles. Operators receive calls from citizens, recording the necessary information on paper forms. Using a map on the wall, the operators attempt to efficiently dispatch emergency vehicles, based on calling the officers via hand held radios. Emergency vehicles must be dispatched within a few minutes of receiving a call, and arrive on the scene within a few minutes.

Emergency Software System for Quick Response currently meets these requirements, and needs a software system developed to streamline the emergency vehicles dispatch process. Management at Emergency Software System for Quick Response also views this as an opportunity to reduce costs (Lauesen, 2002).

- The main purpose of using the Emergency Software System is to allow police and other first responders quicker and more efficient response times to an emergency scene.
- The main beneficiaries of the system are the police personnel, fire personnel, EMS, and citizens who will provide or receive information.
- The main goal is to improve the administration's ability to

collect and interpret data, and to provide real-time image transfers and information sharing system between police vehicles and the command center.

- An emergency software system is developed to alert emergency service personnel quicker and organize incident information.
- Some of these incidents are more serious than others and require a quicker response.
- It is recommended that the system identify incidents based on a system of organization and then allocate resources accordingly.

A citizen calls the hotline for help, the dispatcher types information into the system such as name of person involved and location of the incident. The software sends a GPS signal to available emergency units able to respond based on the input. Then the closest emergency vehicles are dispatched to the scene. Based on the priority and location of the call, the system generates a request for the appropriate number of vehicles to respond to the incident. Details of the incident are logged. As cities have more money available, the emergency vehicles may be equipped with fax terminals. Dispatchers may fax photos or details to the parties involved enroute to the scene of the accident (Chung, Kelly, Moore, & Sanborn, 2006).

2.2. The Background of the Software System

It is critical to have a rapid response system in place for life's unexpected incidents. Emergency software systems for quick response are used to provide information to a designated place within a fraction of time. In the case of an emergency, the emergency software system allows incident report information to be transported immediately to all necessary emergency services personnel. The police department, EMS, fire department, and dispatchers are the users of this system throughout the entire United States.

The software system should be able to provide utility services such as CAD data sharing, GIS/mapping investigation, management, report writing, Scheduling, video analysis, wire tap systems, and other public safety software. The emergency software must consist of SMEAC: Situation, Mission, Execution, Administration, Logistics Command and Communication. The emergency software system consists of the people, software system, and telecommunication devices.

"An emergency response system is comprised of the following:

1. Ontology library for resolving the semantic differences of information pertaining to the incident, its severity, resource requirements, and resource availability
2. Reasoning engine for deriving the resource requirements and emergency response plans based on the policies of different agencies and modes of cooperation
3. Workflow management for visualization, status monitoring, execution, and adaptation of the emergency response process
4. A geographical interface for visualization of situational awareness data" (Niles & Terry, 2004)

The emergency system will be installed at various sites (police stations, fire stations, hospitals, and other emergency control organizations). All of these organizations will be connected by a single network and form a complete structure. Once a citizen places a call or sends an email for help, signals start transmitting across the entire network. This shortens the overall timeframe needed to alert necessary personnel and arrive at the designated scene.

2.3. Step Process of Developing the System Software

There is a five step process in defining the problem in the process of developing the system software: Problem Statement, Root Causes, Stakeholders, System Boundary Diagram and Constraints (Chen & Fu, 1996).

2.3.1. Problem Statement

The problem statement for the manual process was paper based and required multiple copies, manual communication, a slower response time and therefore, delayed the response. The team of architects, project managers, team leaders, software developers and test engineers started by defining the issues that belonged to the particular system before development began. The team developed a software system to analyze the breakdown of the system. In addition, the daily, weekly or monthly reports were produced manually. A manual method cannot generate preformatted reports and analyses of different scenarios. A lack of timing in queueing emergency vehicles resulted in increased wait time and increased severity of injury to citizens. The manual process did not provide proper history and analysis of existing scenarios and data.

The analysis was needed for proper arrangement of future references. The reporting system was needed in order to maintain proper security and alerts. Due to this behavior of the manual process, there are misconceptions between citizens and dispatchers. The citizens are dissatisfied with the manual process. The incident takes longer for emergency responders to arrive on a scene. The benefits of replacing the former manual process with the new software system is that more lives will be saved due to reduced time between placing a call for help and the arrival of an emergency response vehicle. In addition, citizens feel safer when there is a quick response time.

The local governments have separate manual processes for tracking their emergency programs. Many mayors and other elected officials want to switch to software based systems.

It is also decided by the officials and consultants that one centralized system would be the most beneficial. To make things easier, the geographical United States will be split into six regions with one call center representing that entire region. After deciding to implement a software system in place of the manual, the local governments decided to combine the software into one software system for the six geographic regions. This may cause the service to be delayed for each situation. The software will need to be tested for all emergency services.

2.3.2. Root Causes

The emergency system using the fishbone diagram for solving methods can identify the problems. According to Wikipedia, the fishbone diagram visually captured a problem's possible causes and has become a standard in root cause analysis. It begins with a problem, and then identifies possible causes by separate categories that branch off like the bones of a fish. It includes materials, methods, machines, measurements, emergency services and citizens, who can be modified to match a particular issue.

Filing of complaints varies according to each emergency services branch. Before implementing the software system, the manual process to file complaints consists of paper based filing and phone based response. This method can cause a delayed response in the process. Manual processing of registered complaints results in the lack of analysis and reporting. As a result damages accrue, a sense of insecurity in citizens may arise, there would be lost productivity and the system lacks historical analysis and tracking. Figure 2.3.2 Root Causes describes the depth in the Emergency System.

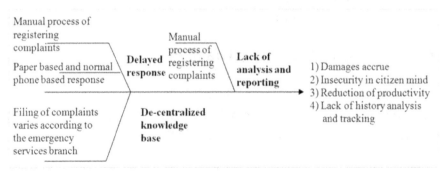

Figure 2.3.2 Fishbone Describe the Depth in the Emergency System

2.3.3 Client, Customer and Stakeholders

According to Kelly Stone of the Collin County, Texas, head of Homeland Security and FBI agent, the police, fire departments, Emergency Medical Services (EMS), administrative staff, dispatchers and citizens are the main groups to benefit from this new system. The Emergency Software System for Quick Response builds the system wishes to meet the goal and reduce spending.

- The police staffs are primary users and stakeholders of the system. They will, upon the request of a citizen, file a complaint and any other related documents for a case.
- The Fire Departments are stakeholders who respond to fire emergencies and backup the police.
- The EMS is the stakeholder who provides care to citizens who are in need of immediate medical attention.
- The administrative staff for the software will be another category of the stakeholders who will be responsible for handling the software system and the need for improvement. They will also be responsible for handling the technical issues, and training additional users of the system.
- The most important customer in the system is the citizen. He or she is responsible for requirement generation, asking for problem support and involvement in day to day requests.
- The client is the local governments build the system wishes to meet the requirements and reduce spending.

2.3.4 System Boundary Diagram

The System Boundary Diagram consists of two factors: the system and the factors outside the system that are affecting it. The figure 2.3.4 discussed some of the components inside the system are below.

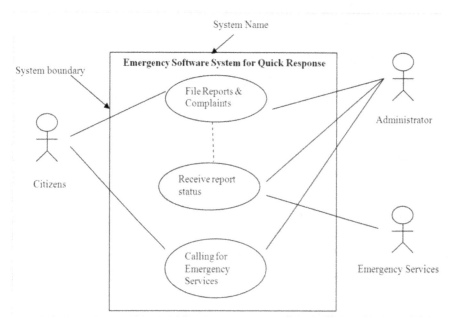

**Figure 2.3.4 System Boundary Diagram
for Inside and Outside Boundary**

2.3.4.1. Inside Boundary

A citizen may file a complaint or report using the emergency software system. They may file the complaint using a phone (emergency phone number) or website. A citizen can register his complaints from the police station, fire department or EMS system. Administrator users can check the reports according to certain parameters, such as daily, weekly, monthly and yearly.

Citizens can track their complaint status via phone or through the emergency services website. The tracking reflects the current status and the actions taken in order to resolve the complaint. Citizens can call the fire/ EMS directly via a landline phone or emergency number box.

2.3.4.2. Outside Boundary

Citizens are responsible for generating complaints and filing the actions needed at the time of the emergency. Citizens are the primary users of the system and responsible for driving the whole system. The emergency services are an important part of system. They are the catalysts in executing the whole process. They are responsible for registering and responding to a primary emergency complaint request. Another important aspect of the outside boundary is the emergency services. Citizens can directly call these services or specific emergency departments. The administrative staff is responsible for tracking system activity and producing accurate reports. The reports reflect the tracking and monitoring of various processed data contained within the system.

2.3.5. The System Constraints

The system constraints specify the limitations that the system has in order to utilize optimum value. The system should produce useable reports based on information from citizens to dispatchers. The expenses should not exceed the budget for the project. This keeps user costs down and leaves extra funds for upgrades, or maintenance, at a later date. The system builds in the following programming languages to write the software, such as C++, Perl, VB, C# or Java, plus an object-oriented language that enables developers to build a variety of secure and robust applications that run on the .net framework. When stakeholders use the emergency software such as C#, a traditional Windows application is created. XML Web services, distributed components, database applications and Graphical User Interface (GUI) make the user interface design friendly. XML will allow citizens to access what services are developed and to evaluate the service from their home computers (Mubarak, 2010).

3

FUNCTIONAL REQUIREMENTS AND TEST PLAN SYSTEM

3.1 Functional Requirements for Use Case

A SWIM LANE diagram illustrates a process workflow which is grouped into three columns. The reason that the swim lane diagram has been selected is because it very clearly illustrates responsibilities of a functional area in a specific workflow (Chung, Kelly, Moore, & Sanborn, 2006). Figure 3.1 illustrates the use case for the swim lane in the diagram below. The workflow for the preauthorization of emergency services consisted of multiple parts involved in the process:

The dispatcher receives a call from a citizen. The emergency vehicle updated the information received from GPS, update emergency vehicles.

A swim lane diagram is grouped into three columns:

Dispatcher Scenario: Receive a call from a citizen and then enter data into the system.

Emergency System: The system will locate the nearest emergency vehicle to the incident by using AVL. This information is displayed for the dispatcher. The system prompts the dispatcher for more information, giving possible suggestions for mistyping or duplicate calls.

The system continues to process a call if it is validated by the team lead. Also the system notifies a team lead if no emergency vehicle is within dispatch range. The team lead then manually calls EMS via landline or cell phone, to go to the scene. Then that information is entered into the system.

Team Lead: Responsible for requesting services, providing services, and appealing denials. Insurer: The insurer is responsible for reviewing the request for services and rendering a decision.

Citizen: The citizen receives the emergency services.

A swim lane diagram for normal scenario according to Wikipedia. org, the most important shape types are as follows:

- Rounded rectangles
  Represent activities;
- Diamonds
  Represent decisions;
- Bars
  Represent the start (split) or end (join) of concurrent activities;
- Black circle
  Represents the start (initial state) of the workflow;
- An encircled black circle
  Represents the end (final state)
- Arrows run from the start towards the end and
  Represent the order in which activities happen.

The software application is used to create a swim lane diagram that will be detailed in the following paragraphs:

1. Dispatcher received call from citizen:

The dispatcher enters the caller's information into the emergency system. The AVL program then signals out to the nearest emergency vehicles and displays these on the dispatcher's screen. The dispatcher can choose the appropriate units and then updates the information into MDT.

1.1. Missing Information:

The system prompts the dispatcher for more information about incident giving possible suggestions using CAD for mistyping addresses or names.

1.2. Multiple calls for same incident:

The system notifies a team lead if there are multiple calls for the same incident. The team lead determines whether the call is to be bypassed by the system. If the call is bypassed, it will be logged into the system and a note that it was not processed will be on file. The system continues to process a call if it is validated by the team lead.

1.3 If the emergency vehicle is not available:

The system will notify the team lead. The team lead calls the next available unit by phone or radio and then updates that process into the system.

2. Emergency vehicle updates the information received from GPS:

A dispatcher retrieves a unit's current position via GPS.

3. Incident status file received:

A dispatcher enters the status file for the Incident.

3.1. Incident requires additional emergency vehicles as necessary:

The dispatch updates the incident information and then enters a new log for each unit incident.

3.2. No incident report is entered after mandated response time:

The systems notify team lead which will determine the necessary update the system, allocating units if needed.

4. Update emergency vehicle information:

The team lead updates the system with changes to the unit.

5. Update location map information:

The team lead will update the system with changes to the city's map.

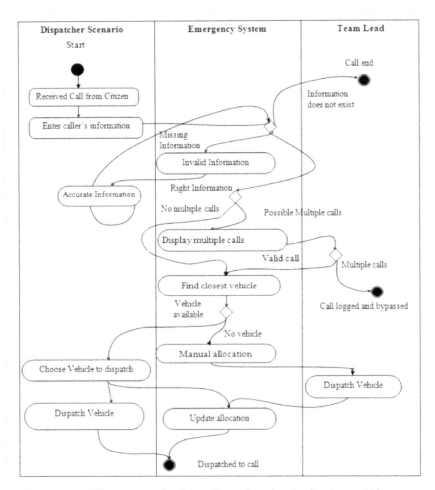

Figure 3.1 Illustrates the Use Case for the Swim Lane Diagram

3.2 Non Functional Requirement

That specifies criteria that can be used to judge the operation of a system, rather than specific functions. While non-functional requirements may eventually map down onto functional requirements, this is generally impractical during the early stages of the requirements process. At this stage, non-functional requirements must generally be documented using informal notations. This makes them intractable to automatic reasoning and thus restricts the applicability of approaches oriented towards the partial automation of inconsistency detection (Boehm, 1988).

Non-functional requirements are difficult to test; therefore, they are usually evaluated subjectively. The following non-functional requirements are responsible for Capacity, Accuracy, Performance Usability, Security and Maintaining the system.

- The capacity of the Emergency Software system for Quick Response depends on the load it can bear at any one time. For example it should be able to handle the peak load during the busiest times. Also it should be able to gather and fulfill the requirements from the area where it has been installed.
- The system should be able to provide accuracy of 100% for a particular incident. For example, it should track the geographical location of the caller while correctly locating a police station for quick help. The system should be capable of handling calls for the fire department and ambulance in a timely manner because delays in this process may cost human lives.
- Performance is the key factor for any particular software. The software works better if it can perform 100% at the time of peak load. The downtime and maintenance time for emergency software system should be kept low in order to provide optimal 24X7 services.
- The usability of the software system lies on the ground. That is how the end user looks for the system. For example the user interface should be simple and interactive. It should provide sufficient guidance to the end user as to how the system should work. If the software system is directly interacting with the voice system then it should provide proper guidelines before anybody chooses an action to perform. There should be proper guidelines and FAQ to guide citizens about the role and working of the emergency software system.
- The software system should have a security aspect as it is related to human life directly. The system should have users who will handle the different area of the software system.

For example the admin user is capable of performing all operations of the system while the operator will be able to read and write the citizen complaints. Network security also plays a key role. For example if the software system is directly connected with the external environment, the main server should have facility to provide antivirus protection for the software system. The antivirus software should be capable of updating itself to prevent the software from new external threats.

- The maintenance factor of the software system is another factor where proper requirement gathering works in long run. For example the maintenance part of the system should be kept low in order to provide a round the clock server. Even for the maintenance portion, there should be a parallel system that takes over if the main system is down. Table 3.2. Produce a List of Non-Functional Requirements for open metric and open target for security.

Table 3.2 Produce a List of Non-Functional Requirements

R1: The software system should be kept under camera surveillance.

R2: The software system should provide enough security to protect it from online and offline threats.

R3: The software system running the computer should be capable of providing the antivirus support.

R4: The software systems should have administrative support with login user name and password.

R5: The number of users accessing the system should be kept low and should be provided with a valid role and credentials.

R6: The security of the whole software system should be 100%.

The metric illustrates requirements for security. For example the requirement R1 shows that the system should be kept in camera surveillance. The R2 requirement provides security via some firewall in the system. The R3 requirements indicate anti-virus support. The R5 and R6 request securing the system by providing valid roles.

3.3 Hierarchical Task Analysis

Hierarchical Task Analysis (HTA) is useful for interface designers in an emergency system because it provides a model for task execution, enabling designers to envision the goals, tasks, subtasks, operations, and plans essential to users' activities (Ellington & Crystal, 2004).

The following diagram is a Hierarchical Task Analysis that addresses one of the main goals of the system, which is demonstrating the process of the Emergency Response System.

The figure 3.3 below explains a High level diagram for Hierarchical Task Analysis.

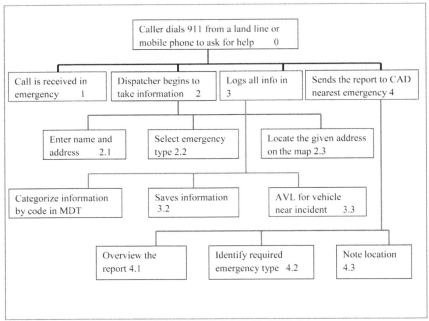

Figure 3.3 High Level Diagrams for Hierarchical Task Analysis

3.4 Instructing and Manipulating

The primary interaction types that will be used in this system are instructing and manipulating.

1) The instructing interaction is as follows:
   » Dispatcher picks options from menus such as the report type, emergency category, and update type.
   » Dispatcher transmits the report to officer(s) via telephone from within the system.
   » Dispatcher clicks the option to return a call to a caller. The dispatcher types information and details about the incident and the caller into the system.

2) The manipulating interaction is as follows:

» Dispatcher is using manipulating interaction when a file is opened and closed or when a file has changed the status of a call. A dispatcher can add and delete details to a call already logged in, just in case it is needed.

» The instructing type is used because the dispatcher will be entering all types of information to create the incident report.

» The incident report is very important when it is being sent to officers in the field. Also, the manipulating type will be used because a call or incident report may be modified or moved, if needed. The file can be modification in case of errors. The report can be moved to a separate site for organizing purposes.

3.5 External Cognition Approaches

The external cognition approaches that will be used to overcome the user limitations are maps, diagrams and drawings. The system would utilize the use of maps and diagrams to help a user do his or her job easier and to eliminate or reduce mistakes due to the nature of the job. Another aid uses electronic reminders of things to do that need a user's attention. A GPS system can be used to coordinate and find incident and emergency locations quicker.

3.6 Perform Usability Testing on First Prototype

Software describes usability focusing on dispatcher testing in six regions. Emergency vehicle command and control system is an example of usability testing for communicating between emergency vehicles command, dispatchers and callers. Experimental design was introduced as a more rigorous form of testing a hypothesis. Nine dispatchers were observed, three per shift for one week. The major problem with the system was defect recording. The dispatchers would save the information under the caller's last name (when known), or assign a number in the name field (when unknown.) The dispatcher received a message stating that the data was saved. However, the information did not show up when a dispatcher would try to locate it by last name or a substitute number. After implementing usability testing principles, it was decided to make changes to the source code. These changes included redirecting the saved information that could be accessed now by name, date, address or event. That provided additional flexibility and ease in moving around the system (Sharp, Preece, & Rogers, 2007).

User interfaces are designed to satisfy usage objectives. This test step uses the designed user interfaces to identify and resolve problems prior to system development. Screen mock-ups are employed to simulate system use. The IT department will identify a standard set of tasks to be simulated. For example, perform a heuristic evaluation on the prototype of tasks to be simulated may include creating, changing, and deleting reservations (Borysowich, 2007).

Perform an evaluation for the user during the test is another task. The IT Department will need to evaluate, constantly talking about what is being done, and what goes into the thought process. A team member will be assigned to take notes on the comments and questions. This test will be based on the user reaction to using the system. This test consists of observing, writing notes and taking questions about experiences and completing interviews. Based on this information, a quantitative and qualitative data analysis can be created using different approaches such as theoretical framework, grounded theory, distributed cognition and activity theory.

Another task is to perform a heuristic evaluation on the prototype visibility of system status. This means that users are kept informed of the system's progress and if they are receiving feedback, GIS will formulate statistical analysis on the information that was collected from the regional centers weekly, monthly and annually in a specified time frame. The project matches up between the system and the real world. By using metaphors involving concrete and familiar ideas, and by making them simple and straightforward, the user can apply his experience and set of expectations to this system by following real-world conventions, making information appear in a natural and logical order. The system should speak the users' language, with words, phrases and concepts familiar to the user, rather than system-oriented terms.

Well-defined, intuitive terminology in error messages, tooltips and documentation will facilitate a better understanding of the system (Lauesen, 2002).

Consistency in the interface allows people to transfer their knowledge and experience from one application or environment to another. Applications can have consistency in many ways. Consistency in the visual design of an application allows the user to learn the visual language of the system more quickly and more confidently. For example, where a dispatch center and vehicle location are implemented consistently, a user who learns what a dispatch center and vehicle location look like and how it functions, will not have to relearn how to make a choice the next time they encounter a dispatch center. A dispatcher will not have to wonder whether words, symbols, situations or actions mean the same thing. Also, the dispatcher will follow well-established platform conventions. Error prevention and good error messages that are informative and helpful to the user are keys to a usable system. But even better is a careful design that prevents the problem from occurring in the first place (Fisher, 2005).

Minimalist design dialogs should not contain information which is irrelevant or rarely needed. Every extra unit of information in a window that is unnecessary or indecipherable competes with the critical information in the window, diminishing its visibility. Do not clutter the screen with too many windows, overload the user with icons, or put dozens of buttons in a dialog box. Keep gratuitous visual clutter to a minimum and ensure that the graphics that are presented are of top quality and legibility (Lamsweerde V. , 2009).

The system is simple as possible, so as to allow fast dispatch, while meeting the information distribution and organization requirements. The interface will be GUI based. There will be a login window, a data entry window, a status window and a statistics window. The data entry window will be used by dispatchers and the team lead to input information. The status window will display errors to the team lead for resolution. It will also display all active calls. The statistics window will display information such as average response time.

3.7 The Test Log

The test log was checked by heuristic evaluation to ensure that everything is set up correctly. A dispatcher called the system to confirm that all tools worked well together, and then the project proceeded as usual. The deliverables are measured to meet the test's objectives.

The activities are measured against the test deliverables production. The system is tested weekly to find out what defects are in the system. Written documentation tracked all the steps from issue to completion throughout the entire process. For the software a test log was run every Sunday at midnight for one year. A few issues surfaced. Based on a heuristic evaluation, the problems that were identified and resolved are as follows:

A. Defect Following
   1. Defect Recording - Describe and quantify changes from requirements
   2. Defect Reporting - Defects, including severity and location
   3. Defect Tracking - Defects from the time of recording until a conclusion has been reached

B. Testing Defect Correction
   1. Validation - Changed code and documentation at the end of the change process to ensure that software requirements were met
   2. Regression Testing - Testing the product to see that unchanged functionality performs as it did prior to implementing a change
   3. Verification - Reviewing requirements, design and documentation to make sure that they are updated correctly as a result of a defect correction

3.8 Overview Prototype of the User Interface

The overview for a prototype is a representation of the dispatcher interface that is built early in the development cycle with the express purpose of being changed and improved. The purpose of such a prototype includes training users on the system before it is released and then improving the dispatcher interface. The dispatcher interface allows the employees in the administrative department to enter a citizen's information and to request help. The recommended approach for development is to define the prototype objectives, choose the prototyping tool, build and investigate the prototype. The validation and verification process is to ensure that the user interface prototype will be good enough to enable user interaction. Then the programmers will gather useful feedback, detailed and complete enough to support evaluation based on available requirements analysis.

Some of the feedback indicated a need for the interface to be more user friendly. To solve that issue, menu bars have been added. The menu bar contains different menus with common commands to perform different functionalities on spatial data. The map toolbar has buttons including visual interaction, querying and an output control. Menus, submenus and the tool bar are functional descriptions. Using a set of Virtual Windows (VW) in a system involved use case scenarios to identify the error condition that needed to be supported. Use worst case data values to show how the VWs were accommodated (Sharp, Preece, & Rogers, 2007).

The login window contains the login facility for the user in figure 3.8a Login Window. There are different kinds of users that exist in the system, such as normal or administrator. The user function provides a name and password. The system grants rights to the user to access various areas of the system such as the reports and operations.

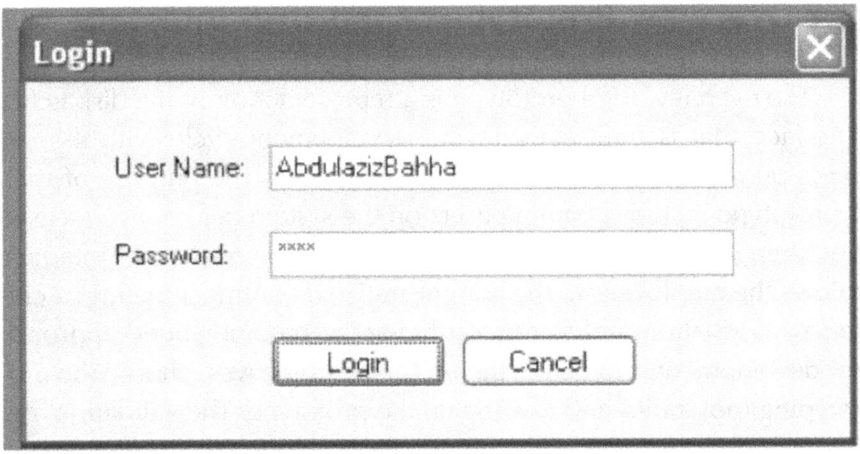

Figure 3.8a Login Window

The main user interface depicts the way the window will appear in figure 3.8b in the system. There will be a main window that will work as a container. The user interface window function to the screens for identity. The menu and status bar will be a part of the container window. There will be different options for the user in the menu.

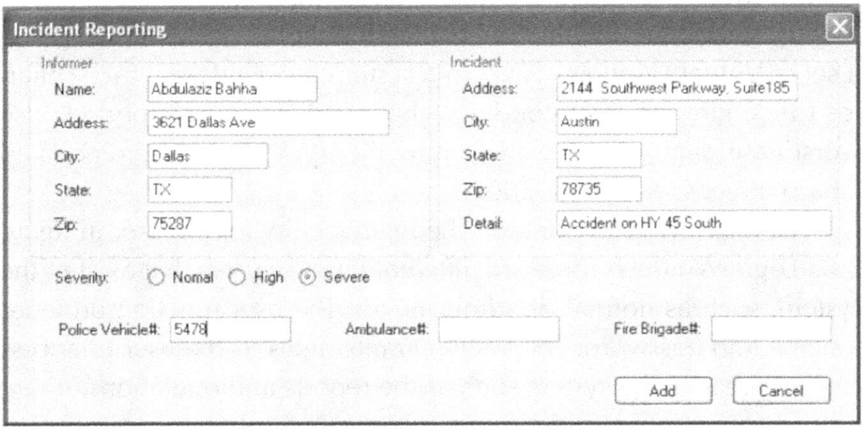

Figure 3.8b Main User Interface Window

Tracking the main window contains information regarding the incident and the related events in Figure 3.8c Incident Tracking Window. The window contains the caller information and call details such as name, address and time of call. The right hand side window contains the detail regarding the incident. The report generation window contains the main report category and sub report category. The main report categorizes information daily, weekly, monthly and yearly. Selecting the main category prompts the sub category to pop up. Once clicked, the report will be generated at the bottom.

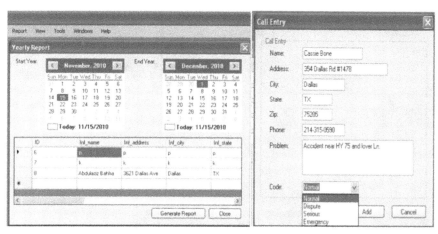

Figure 3.8c Incident Tracking Window

4

DATA AND PROCESS REQUIREMENTS GATHERING

4.1 Data Requirements

A DATA REQUIREMENTS gathering is a vital part of successful emergency software and tools development. The Emergency Software System will streamline the emergency services software system. The Emergency Software System proposes to divide the United States into six regions. Each region will have an emergency services center. All calls from citizens will be routed to the nearest regional center. A dispatcher that receives a citizen's call enters the information into the system and requests the emergency vehicles that are needed at the scene. Streamlining this process will cut costs for city and town budgets. Also, the increased efficiency will cut down on wasted time and resources.

According to Robert Brown, of the Missouri University of Science & Technology, in the United States, more than 156,000 accidents involving emergency vehicles occurred at intersections from early 1980s to 1995, resulting in 6,550 deaths. National Safety Board statistics show that 40% of firefighters killed in the line of duty died in accidents on the way to a scene, due to slow response time and miscommunication of map locations or databases for the emergency vehicles. Additionally, the interface was unorganized and the internet was not available before the early 2000s. These tragedies may have been prevented by using the Emergency Software System for Quick Response.

The Emergency Software System for Quick Response will solve the problem by providing an early warning of approaching emergency vehicles. There are three main components for data requirements: map locations of a citizen's physical location, databases of citizens' addresses, and the interface tools that allow planners to integrate these features. These components help a dispatcher to share information in real time without waiting for a response from emergency vehicles.

The Emergency Software System tools development is a planning toolkit, which allows the emergency services to query and examine data, create and retrieve emergency response plans and generate maps and reports. Other tools are for database maintenance, including tools to edit street addresses, look up supporting information efficiently, and maintain the supporting database tables. The GIS tool based planning system is designed for the emergency planners. They address emergency management issues, such as capabilities, to support the collection. Editing tools for the addressed street layer was a major issue before the Emergency Software System for Quick Response was developed and implemented.

The problem with the former method of editing tools was that many of the tool elements, such as a list of invalid street type abbreviations and misspelled cities, were loaded from text configuration files that were easily modified by the dispatcher or other emergency services personnel. Allowing multiple employees access to the text files to change them could have resulted in mismanagement, miscommunication and harm. Another challenge was that the former method tracked incidents by using three layers of carbon copy sheets. Technology became more accessible over time.

Now, Emergency Software for Quick Response solves those problems using Geodatabase tools. Also, GIS tools give the dispatcher access to coordinates and attributes of point layers stored in the database. GIS event themes are used as the means of converting a table with coordinates to a map layer. The system also provides alternate methods of locating points for cases when the Geodatabase fails, or the emergency service wants to improve the accuracy of the location. The emergency service system takes advantage of standard GIS Geodatabase functionality and augments it with a procedure to store the results in fuzzy logic and display the locations as a GIS event theme.

Another data gathering tool is fuzzy logic. According to Wikipedia, fuzzy logic starts with and builds on a set of emergency system user supplied developers' rules. Additional benefits of fuzzy logic in the Emergency Software System include its simplicity and its flexibility such as handling problems with imprecise addresses or other incomplete data. It can model nonlinear functions of arbitrary complexity.

CAD system maps were provided by several states and are beneficial for updating street names and address ranges. When a CAD system is provided, they are used to validate the information in GIS. When these materials are not provided, or are not available, commercial maps, road databases and web based mapping programs are used to help identify street names and address ranges.

AVL is the technique of using a navigation system, such as GPS, to determine a vehicle's position. AVL improves navigation guidance and produces better time estimates and route optimization. An independent AVL created for a single route meets specific needs. It is able to export real-time data to a central server for real-time use. The AVL will include displays on exterior visual screens for at least 6 to 10 emergency vehicles including emergency services station locations. It also reports the next emergency vehicle's time arrival visually and in English or Spanish. The AVL system must include the ability to send the following data in real- time to a regional server in XML or another chosen system software language (Portillo, 2008).

4.2 Former Traditional Manual Process and New Emergency Software System

The Emergency Software System is the software written and built to increase the manual process and decrease the time needed to potentially save lives. When a citizen placed an emergency call for help, the manual process caused delay because information was hand written, then a dispatcher had to locate the appropriate emergency service vehicle, send a help request over the radio, wait for a response, then radio back the request for help. This former process was very time consuming and caused more suffering for the severely injured citizens.

The technologically advanced Emergency Software System has the following benefits over the manual process of providing information:

1. It is faster and has nearly zero probability of error. It is more efficient than the manual process.
2. It is a computerized system of combining the three sub systems (i.e. police, fire, and EMS.)
3. The former process required the dispatcher to place separate radio or phone requests to each department. For the technologically advanced Emergency Software System, a citizen has to call only a single help line number and based on the situation the system alerts all the sub systems to follow the guidelines.
4. The initial investment in the software is a one time process. That cost may be depreciated over several years. This cost is cheaper compared to the labor cost included in the manual process.
5. The Emergency Software System provides quick access to all stored data in order for administrative staff to run multiple reports that retrieve queried data. This benefits managers in making informed decisions in a short time period.
6. In summary the system is a new technology based solution that increases the flow of information in times of disasters and emergencies.

4.3 Design Constraints

Constraint diagrams are used for optimization. The emergency software system is using C# as the language for development. C# is the best choice for this software because:

1. It is a managed language that needs Microsoft Visual Web Developer 2010 Express Edition (Microsoft .net) run time to execute the program. That means that the run time manages the garbage collection, memory management and other services instead of the language. This decreases the amount of time developers have to spend coding.
2. The C# language is dependent on the .net run time. It only requiring the run time to exist in the system to execute the program.
3. C# may be installed in various platforms. C# has a strong type casting mechanism that makes it a malleable language for development.
4. C# allows support data file access for administrative staff and dispatchers.

4.4 Implementation Constraint

The figure 4.4 Implementation Design Constraints below describes how the design constraints work for a telephone system call between a citizen, a dispatcher and the appropriate emergency service. A caller places a call for help from a cell phone to emergency services. The signal appears on a dispatcher's screen in the form of an address. The dispatcher will ask the caller questions about the incident. The dispatcher will send a message via radio communication to an emergency vehicle. The dispatcher will see which emergency vehicle is available using an AVL system that is designed to address emergency response and preparedness. Mobile data terminals (MDT) store and disseminate information. The vehicle location is transmitted to the dispatcher via GPS to towers located in the downtown areas for each region (Chung, Kelly, Moore, & Sanborn, 2006).

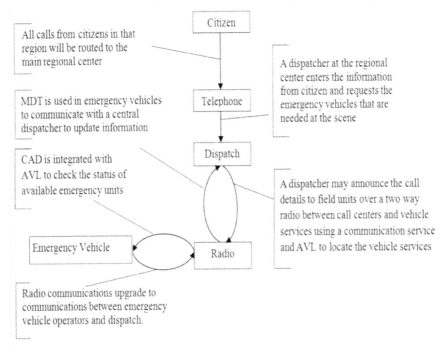

Figure 4.4 Implementation Design Constraints

The important shape types are defined as follows:

Data store

Line with an arrow indicate the direction of the data flow

Rectangle indicates an activity expanded to span multiple lanes if needed

4.5 Software Applications

Several other software packages are used to support system and database development. Table 4.5 summarizes these software packages. The Microsoft Visual Web Developer 2010 Express Edition and Operating System are utilized by programmers, or front users, to develop the emergency system.

Software Application (Office 2007) Photoshop and Text editor are utilized by end users such as dispatchers and other administrative staff to enter data, run reports, and request emergency services.

Table 4.5 Summarizes the Software Components Used for Development

Type	Product	Use	Description
Software Application (Office 2007)	Microsoft	Word, Access, Spreadsheets	end users can create working applications
Microsoft Visual Web Developer 2010 Express Edition	Microsoft	C# Program Language	front users develop the emergency system
Operating System	Microsoft	Windows Vista	front user program that pulls the system together
Photoshop	Adobe	Interface Icons and Figures	end user application to enhance 3D images and motion-based content
Text editor	IBM Corporation	Extensible Markup Language (XML) for Web services	end user application to create Web Page

4.6 Use Cases for System Functioning Requirements Scenario

The main purpose of using the context diagram is that the background, or settings, determine, specify, or clarify the meaning of an event. The emergency software system allows emergency services more efficient response times to an incident scene. The Context diagram in figure 4.6 below represents the actors outside the system that could interact with that emergency system (Chung, Kelly, Moore, & Sanborn, 2006). The main beneficiaries of the system are the police station workers, fire station workers, EMS (Emergency Service), and citizens who will provide or receive information.

The main goal is to improve the administration's ability to collect and interpret data provide real-time image transfers, and share information system between emergency vehicles and the command center.

An emergency software system is developed to alert emergency service personnel quickly and to organize incident information. It is recommended that the system identify incidents based on a system of organization then allocate resources accordingly.

There are four types of entities available that interact with the system.

- A citizen calls in a request to dispatch to send help for an incident involving bodily injury. Dispatch forwards the request to the emergency services to go out to the scene.
- The dispatchers receive incoming calls and route them to the appropriate departments (i.e., fire, police, and EMS.) The dispatch operators are the gatekeepers of the system.
- Emergency services are an important entity in interacting with the central system. Emergency services receive dispatchers' AVL request.
- The administrator is mainly responsible for getting reports and checking for the complaints that have been generated by the citizen. The administrator provides parameters in order to generate the reports. Once generated, they analyze the out-

come and make decisions regarding the whole system.

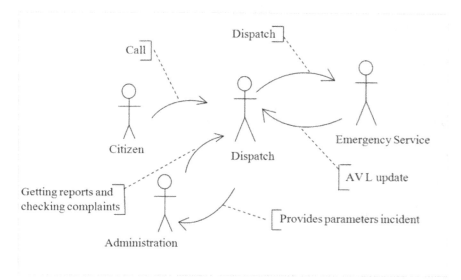

**Figure 4.6 The Context Represents the Actors
Interacting with the System**

4.7 Risks in Development

There are risks in the development of an Emergency Software System. The risks include time delays, GPS issues, undereducated staff, complexity of the software, and survivability.

Time delays are due to operator error. Also, former technology, delays the overall quality of response. To mitigate this situation, employees should receive training and continuing education.

Normally, the dispatchers rely on the accuracy of GPS to direct the emergency services personnel to an accident scene. However, there is a risk that the GPS directions will not match the conditions on the way to the scene. The variance is almost always due to local construction that wasn't picked up by the satellite because GPS wasn't updated. The way to mitigate this risk would be to receive local feeds that reflect the current construction process for a given area.

The staff members who will work with the system day to day lack technical knowledge. This lack of knowledge is the greatest risk in the development phase. If the staff is undereducated about technology, they may not be able to tackle the situation where human interaction is required. For example, while functioning, if the software system stops working then non technical staff will be not be able to recover the system instantly and may delay emergency services from arriving at an accident scene in a timely manner. To remedy this situation employees should receive training and continuing education.

The complexity of the software is another risk in the development of an emergency software system. The complexity involves a scenario that has been handled in the software system. For example, software has been made for tackling the predefined number of delayed responses that have resulted in deaths. If the number of failed responses continues to increase then the software fails. So it is better to minimize the number of cases and take the practical scenario in mind to design emergency software. The complexity lies with the formation of the cases and algorithm. The quality of the algorithm will decide how the system will behave in a complex situation. To maintain this quality the programmers should ascertain the appropriateness of the step complexity (SC) measure. This is done by averaged step performance time data obtained from emergency training records that shows for the loss of coolant accident and the excess steam dump per emergency incident with estimated complexity scores.

Survivability is another potential risk for developing the emergency system. It is the property by which any system works properly at the time of attack, failure or accident. To ensure survivability, it is necessary to have proper analysis and attention towards natural climate.

5

DIAGRAM MODELS OF THE PROPOSED SYSTEM

5.1 Sequence Diagrams Overview

EMERGENCY SOFTWARE SYSTEM used a sequence diagram to show the interactions between objects in the sequential order where those interactions occurred. Sequence diagrams consist of rectangles placed on top of the page horizontally. The rectangles represent the classes that interact in the scenario. Vertical lines coming out of this rectangle are actors called lifelines.

Each Lifeline element represents the life of a given object. Lifelines are connected by horizontal lines denoting messages that pass from one object in the scenario to the next object in the scenario. A sequence diagram captures the behavior of the scenario. The diagram shows a number of objects and the messages that are passed between these objects within the scenario (Basterra, 2010).The system uses sequence diagrams to explain the behavior of several objects in a single use case, such as the object creation and deletion, and the complex sequences of calls and the return values.

5.2 The Sequence Diagram for the Normal Scenario

According to Wikipedia the sequence diagram for the normal scenario shows how the citizen receives a response from the system in an ideal setting. The citizen calls the emergency system in order to get help. The system determines the severity of the request and calls the police or fire or EMS. The request is logged and the response is provided to the user. The information is shown by flowing into the system by the arrow. The dotted lines indicate the boundary of an entity. The caller represents an external person that interacts with the system, and the line with an arrow indicates the direction of the data flow. Figure 5.1 is a sequence diagram that shows one implementation of that normal scenario. Sequence diagrams show the interaction by showing each participant with a lifeline that runs vertically down the page and the ordering of messages by reading down the page (Stair & George, 2008).

Each lifeline has an activation bar that shows when the participant is active in the interaction. Also, the system used one return line. However, the system could use them for all calls. The first message doesn't have a participant that sent it, as it comes from an undetermined source. Sequence diagrams clearly show how objects interact. They aren't good at showing details of algorithms, such as loops and conditional behavior, but they make the calls between objects clear and give a really good picture about which objects are doing which processing.

The objects in this Sequence Diagram have a more distributed control. This is good since it is more suitable for polymorphism.

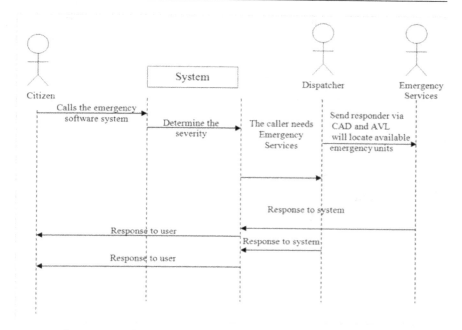

Figure 5.2 Show Sequence Diagrams Normal Scenario

The important shape types are defined:
Vertical boxes:
Activation frames Solid horizontal arrows: request messages (usually function/method calls)
Arrow points from caller to object called
Label: name of function being called
Includes parameter names, types, if this clarifies call
Dashed vertical arrows: response messages (usually method/function call return values)
Label: description of value being returned
Column headings can also be used
Case actors requesting actions Classes (to indicate static methods)

5.3 Sequence Diagrams Abnormal Scenario

The diagram demonstrates the system's behavior when an abnormal scenario is in play. When the system is overloaded with requests

or there is a fault in the software and hardware, then an abnormal scenario results. In the figure 5.3 the abnormal scenario diagram shows
how the system connects with the police, EMS and fire stations. If
the system is busy, overloaded or down because of maintenance, the
dispatchers will call the units by radio. In the abnormal scenario,
the system works in order to provide assistance to the end user. A
common issue with sequence diagrams is how to show looping and
conditional behavior. Here is the notation to use: both loops and
conditionals use interaction frames, which are ways of marking off a
piece of a sequence diagram. Figure 5.3 is based on the response to
the user regarding overloading.

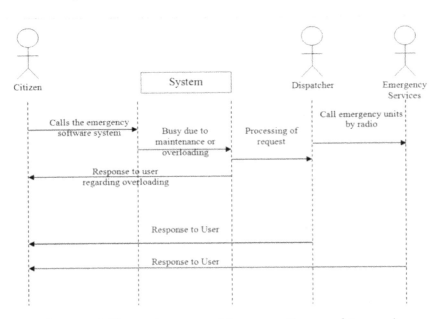

Figure 5.3 Shows Sequence Diagrams Abnormal Scenario

5.4 Representative Use Cases System Functioning

The Use Case Diagram has external entities working on the system.
Those are citizens, administrators, police, fire department and EMS.
They interact with the system to perform different tasks. Then emergen-

cy services are alerted when the incident needs assistance from those areas. The administrator staff member is responsible for providing parameters that generate the reports. Figure 5.4.3, a set of representative use cases for the system (System Functioning) for Whole system Use Cases for System Functioning Normal and Abnormal and Misuse Scenario. The emergency software system has a piece of behavior that is similar across many use cases. Administrative Staff break this out as a separate use case and let the other ones "include" it. The system include: valuation, validate user interaction and check for proper authorization (Taylor & Medvidovic, 2010).

5.4.1 Use Cases for System Functioning Normal Scenario

The purpose of the emergency alert for incident system is to provide timely notification and warning to all citizens of a threat, occurring or imminent, that poses an immediate threat to their health, or safety. The Administrative Staff is responsible for providing immediate emergency response and investigating reports. The following will explain each part of Alert Incident and Administrative Reporting.

Alert for incident allows for the integration of facilities, equipment, personnel, procedures, and communications operating within a common organizational structure. It also enables a coordinated response among various jurisdictions and functional agencies, both public and private. In addition, it establishes common processes for planning and managing resources.

- Pre-Condition- A citizen report for a particular incident that has occurred somewhere in the city.
- Post-Condition-The system will respond to a citizen's and call, alert the nearest police station, fire, and/or EMS facility.
- Description-The citizen will call the emergency number given for the appropriate emergency system.

Once a citizen is connected with the response system, it will calculate the severity of the incident and call the corresponding police station.

If the incident needs the fire department to be dispatched, it will request a response from the nearest fire station. If the victim needs assistance from the ambulance services, it will redirect the request from the system to the nearest EMS facility.

Administrative Reporting is responsible for receiving, evaluating, and analyzing all emergency related information and providing updated status reports to emergency operations plan management and field operations, for the action planning function within the providing, in conjunction with emergency operations plan management, concise overview and direction for each operational period.

- Pre-Condition – Any member of an administrative staff may request reports related to any incident.
- Post-Condition – The system will generate the report based on the criteria (can be exported in various format). Security measures will be installed and enforced.
- Description – Administrative staff will log into the system with secured passwords and various options will appear on the screen.

Once a staff member selects a particular report, he/she will be asked to enter options for various parameter values. The users are given the option to export the report in different formats such as Document, CSV file, and XML.

5.4.2 Use Cases for System Functioning Abnormal Scenario

The purpose of using Abnormal Scenario is to compare the new interfaces to the current interface across normal and abnormal scenarios. There are a number of different factors that go into the design of emergency and abnormal checklists. How these factors may relate to checklist use and errors made when completing them is often under-appreciated. A variety of checklist design features are discussed with an emphasis on how they influenced the system. There are two Scenarios for System Functioning Abnormal Fault due to overloading and system downtime (Alexander, 2003).

Scenarios for System Functioning Abnormal Fault due to overloading are:

- Pre-Condition – The system starts to work abnormally or slow down in response, when overloaded with the resources or incident calls.
- Post-Condition – Switching to a manual system, or alternate backup system is an option for responding to emergency calls.
- Description – The emergency software system can be overloaded with incoming calls and may not work under the planned conditions.

To overcome this situation there are two solutions. The first solution is to provide a backup system that will work parallel to the original system and start functioning once the first system gets overloaded. A second solution to the problem is to provide a manual system alternate to the system that becomes overloaded.

Scenarios for System Functioning Abnormal System downtime:

- Pre-Condition – Every system needs maintenance tasks performed at regular intervals of time. The period of time when maintenance work is transpiring is called downtime.
- Post-Condition – All system requests should be transferred from the alternate system which is covering for the main system when it is in downtime.
- Description – Due to lengthy maintenance tasks, the system may be in downtime for several hours. This time is utilized by the technician to check for faults in the system and provide the new fixes, so, during maintenance, the parallel system picks up the slack in order to provide a non- disturbed facility.

5.4.3 Use Cases for System Functioning Misuse Scenario

A Misuse Case is simply a Use Case from the point of view of an Actor hostile to the system under design. Misuse Cases turn out to have many possible applications, and to interact with Use Cases in interesting and helpful ways. Like any other communications network, emergency communications can become for sure the target of misuse and attacks. There are some special characteristics of the emergency services which allow the emergence of new kinds of attacks that are not visible in other communications networks (Alexander, 2003).

These characteristics are:

- Pre-Condition – The systems users (citizen) do not have proper knowledge about the emergency system. The system may receive improper calls that are not related to any emergency.
- Post-Condition – The system should get an interpreter to determine whether the calls are true emergencies. If not, proper assistance will be provided to the citizen regarding the uses of the system.
- Description-Due to unavailability and improper guidance regarding the system, it could be easy to misuse the system. For example, a citizen may call the emergency line to lodge a complaint, which is an improper use of the system.

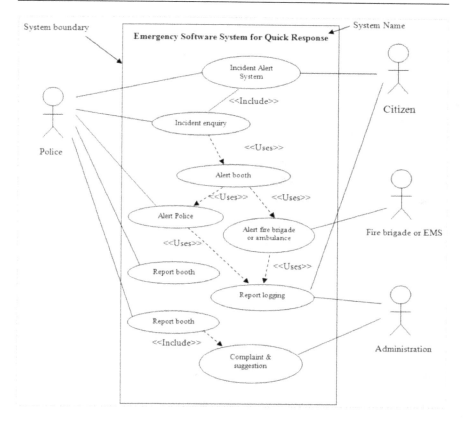

Figure 5.4.3 Systems Functioning for Whole System

6

SUMMARY AND CONCLUSIONS

6.1. Introduction

IT IS A requirement of emergency service that it responds as quickly as possible to reported incidents. The objective of a command and control system is to ensure that incidents are logged and routed to the most appropriate emergency vehicle. Factors which must be taken into account in deciding which vehicle to send to an incident include:

The types of incident-some incidents are more serious than others and require a more urgent response. It is recommended that classes of response be identified and incidents allocated to these classes.

Tools are designed for Emergency Software for Quick Response used in database and interface design will allow it to be useful for other planning activities in the organizations receiving it. The need for a county-level emergency planning system with detailed, geographically referenced information about populations, facilities, potential events, resources, and control points is not specific to this hazard, or this location. The data collection methods, database content and design, geocoding methods and improvements, interface design, and the approaches to documentation and training are all broadly applicable. The use of GIS to support emergency management, both in response activities and planning, is increasing. As this project demonstrates, developing such a system for a localized area is both uniquely valuable and uniquely challenging.

The position of available vehicles-the best strategy is to send the closest vehicle to respond to an incident. Take into account that the position of vehicles may not be known exactly and it may be necessary to send a message to vehicles to determine their current position. The type of vehicles available-some incidents require a number of vehicles. Others such as traffic accidents may require specialized vehicles. The location of the incident-it may be unwise to respond to an incident by sending a single vehicle. In other areas, a single vehicle or policeman may be all that is needed to respond to the same type of incident. The need to alert other emergency services such as fire and EMS, the system should automatically alert these services if necessary. A dispatcher will log the incident report details into the system. A system such as this one is open to indefinite expansion.

Emergency vehicle command and control system requires the officer to activate the warning and recording equipment and report to the command control center upon engaging in a pursuit. It is difficult to perform all of these actions at the same time, so this project provides an automated, integrated solution. In order to function efficiently the command control center, dispatch has to meet requirements to work with the system: wireless internet, vehicle information automated pursuit report, global wireless network, vehicle diagnostic information, automated voice alert and auto tracking pan and tilt camera. The product life cycle for the emergency system completed using the interaction software development model.

6.2. Future Work

The next generation work will look towards improving mobile computing support for professionals involved in a disaster response and recovery operation to facilitate better assessment of the damage caused to life or buildings and to make this assessment information available to personnel (dispatchers or employee) within the disaster response arena so as to expedite a safe, efficient and effective disaster response process. The research method involved the use of scenario-based emergency services needs analysis for studying end-user needs and requirements and use of rational unified process for software design and implementation. IT support collaboration platform is developed to enable first responders to communicate using handheld devices and laptops and share critical building evaluation information using a mobile ad hoc network. Radio Frequency Identification (RFID) enables mobile devices and tags to be used for posting, gathering, storing and sharing building assessment information in an efficient manner with fewer errors, leading to improved efficiency and effectiveness in the emergency response process (Bissell, 2009). The key research contribution includes analysis of the information needs of first responders, development of a collaborative framework for supporting urban preparedness and emergency response, demonstration of developed concepts in realistic disaster scenarios, and implementation and validation of the prototype system to demonstrate the concepts. Based on the composition approach, flexible information systems can be build which are adaptable to special use cases by appropriate composition of components. This allows for the same system to be used over a product lifecycle or for multiple phases of the emergency management cycle. The prototype presented covers only a small fraction of the potential of this approach. Work is under way to incorporate new functionality (like geographic information components, GIS) and exploit dynamic procedures.

In the future, this approach must take the following:

Using a mobile ad hoc network Interactive Voice Response (IVR) systems the citizen can call to date help information instantly automates interactions with telephone without having to speak directly to a person existing systems and tools must be able to interface incorporate new functionality geographic information components, GIS exploit dynamic procedures

Continue developing one all-purpose system for communication maintenance and system enhancements should be cost effective IVR platforms may also offer:

The ability to recognize spoken input from callers (voice recognition), translate text into spoken output for callers (text-to-speech) and Transfer IVR calls to any telephone or call center agent.

6.3. Glossary

Abbreviations	Meaning
AVL	Automatic Vehicle Location
GPS	Global Positioning System
CAD	Computer Aided Dispatch
GIS	Geographic Information System
SMEAC	Situation, Mission, Execution, Administration and Logistics, Command and Signal
MDT	Mobile Data Terminal
IVR	Interactive Voice Response
SC	Step Complexity
ERD	Entity Relationship Diagram
GUI	Graphics User Interface
EOP	Emergency Operations Plan
XML	Extensible Markup Language
DFD	Data flow diagrams
CASE tool	Computer-Aided Software Engineering
SRD	System Requirements Document
CSV	comma-separated values
SS	System Specification
SMS	Short Message Service
VW	Virtual Windows
IVR	Interactive Voice Response
RFID	Radio Frequency Identification

BIBLIOGRAPHY

Alexander, I. (2003, January). *Use Cases with Hostile Intent*. Retrieved 8 2010, from http://easyweb.easynet.co.uk/~iany/consultancy/misuse_cases_hostile_intent/misuse_cas es_hostile_intent.htm

Basterra, L. (2010, August). *How to create good Sequence Diagrams*. Retrieved 11 2010, from bell south: http://coweb.cc.gatech.edu/cs2340/6403

Bissell, R. (2009, June). *Research Digest*. Retrieved from Natural Hazards Center:

http://www.colorado.edu/hazards/rd/rd_jun2009.pdf

Boehm, B. (1988). *A Spiral Model of Software Development and Enhancement*. Retrieved 9 21, 2010, from IEEE.

Booch, J. R. (2004). The Unified Modeling Language Reference Manual. In I. Booch. Addison- Wesley.

Borysowich, C. (2007, September). *Usability Testing*. Retrieved 7 2010, from Information Technology Toolbox, : http://blogs.ittool-box.com/print.asp?i=18788

Carpenter. (1999, 10). Verification of Requirements for Safety-Critical Software. Redondo Beach, CA, USA. Chen, & Fu. (1996).

Chung, L., Kelly, R., Moore, S., & Sanborn, J. (2006, 2). *ambulance dispatch system*. Retrieved 6 2010, from utdallas.edu: www.ut-dallas.edu/~chung/CS6354/...U07.../Deliverable_0_Fantastic9.doc

Easterbrook, S. (1993). *Domain Modelling of Hierarchies of Alternative*

Viewpoints. Retrieved 9 8, 2010, from 1st IEEE International Symposium on Requirements Engineering, San Diego,.

Ellington, B., & Crystal, A. (2004). Task analysis and human-computer interaction. *New York.* New York: Proceedings of the Tenth Americas Conference on Information Systems.

ESRI. (2005). *GIS Software.* Retrieved 6 2010, from Mapping the Future of Public Safety: http://www.esri.com/library/brochures/pdfs/public-safety.pdf

Fisher, M. (2005). Mobile Data and Automatic Vehicle Locations.

FTA. (2007, 4). Retrieved 9 2010, from http://www.unitedweride.gov/Cost_Allocation%281%29.pdf

FTA. (2007, 4). Retrieved 5 2010, from http://www.unitedweride.gov/

Guo, F., Ji, Y., & Hu, G. (2002). *GIS Development Magazine*, Methods for Improving the Accuracy and Reliability of Vehicle-borne GPS Intelligence Navigation.

Guo, F., Ji, Y., & Hu, G. (November 2002.). Methods for Improving the Accuracy and Reliability of Vehicle-borne GPS Intelligence Navigation. *GIS Development Magazine.*

Jacobson, I. (1999). *The Unified Software Development Process .* Boston: Pearson Education.

Käkölä, i., & Dueñas, J. (2006). *Software product lines: research issues in engineering and management.* New York: springer.

Kuiper, J. A. (2000). DEVELOPMENT OF A GIS-BASED EMERGENCY PLANNING SYSTEM. Argonne: Twentieth Annual ESRI User Conference.

Lamsweerde, A. V. (2009). *John Wiley & Sons Inc.* Hoboken: John Wiley & Sons Inc.

Lamsweerde, V. (2009). Requirements Engineering: From System Goals to UML Models to Software Specifications. Hoboken: John Wiley & Sons Inc.

Lauesen. (2002). *Software Requirements: Styles and Techniques.* Addison-Wesley.

Lauesen, S. (2002). *Software requirements: styles and techniques.* Addison-Wesley.

Malan, R., & Bredemeyer, D. (2001, August 3). *White paper .* Retrieved 8 5, 2010, from Bredemeyer Consulting : http://www.bredemeyer.

com/pdf_files/functreq.pdf

Marsh, S. (2007). *Risk and business continuity management*. Retrieved 2010, from the Business Continuity: http://www.bcipartnership. com/businesscontinuitymanagementguide0809.pdf

Mosavi, M., & Ghadiri, A. (2002). *Implementation of a GPS-based AVL System for Mazandaran Province*. Retrieved 9 2010, from GIS development: http://www.gisdevelopment.net/technology/ mobilemapping/ma05211pf.htm

Mubarak, S. (2010). *Construction Project Scheduling and Control*. Wiley, John & Sons.

Niles, I. T., & Terry, A. (2004). The MILO: .A general-purpose, mid-level ontology. *International conference on information and knowledge engineering* (IKE'04).

Portillo, D. (2008, February). *Automated Vehicle Location using Global Positioning Systems for First Responders*. Retrieved 3 2010, from http://www.usafa.edu/df/iita/Technical%20 Reports/Automated%20Vehicle%20Locator%20Using%20 Geoposition%20Positioning%20System%20for%20First%20 Responders. pdf

Sawyer, P., Sommerville, I., & Viller, S. (1996). *Tackling the Real Concerns of Requirements Engineering*. Retrieved 9 30, 2010

Sharp, H., Preece, J., & Rogers, Y. (2007). *Interaction design: beyond human-computer interaction*. Hoboken: Wiley.

Sims, R. (1997). *"Interactivity: A forgotten art?" in Computers in Human Behavior*. Retrieved 11 2010, from http://www2.gsu. edu/~wwwitr/docs/interact/

tair, R., & George, R. (2008). *Fundamentals of Information Systems*. Cengage Learning.

Taylor, R. N., & Medvidovic, N. (2010). *Software Architecture: Foundations, Theory, and Practice*. Hoboken: Jone Wiley& Sons.

Testing, S. (n.d). Retrieved 3 2010, from smart software testing: http:// www.smartsoftwaretesting.com/edu/CSTE/

Waldman, S. (1999). *GISdevelopment*. Retrieved 9 2010, from Mobile Data Technology at Florida Power Corporation: http://gisdevelopment.net/proceedings/gita/2001/mobile/mobile005pf.htm